STARK LIBRARY NOV - - 2022

DISCARD

EXPLORING THE ELEMENTS

Nickel

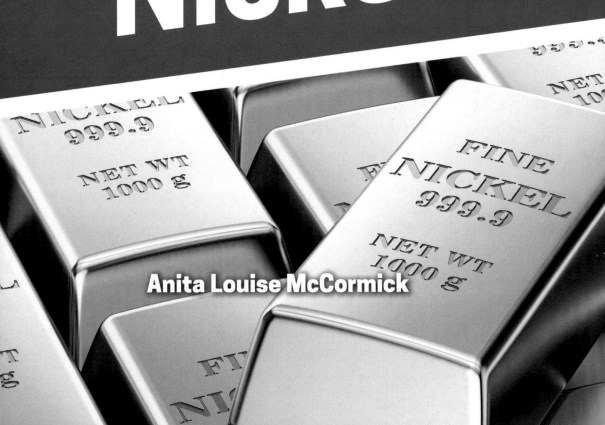

Anita Louise McCormick

Enslow Publishing
101 W. 23rd Street
Suite 240
New York, NY 10011
USA
enslow.com

Published in 2019 by Enslow Publishing, LLC.
101 W. 23rd Street, Suite 240, New York, NY 10011

Copyright © 2019 by Enslow Publishing, LLC.

All rights reserved

No part of this book may be reproduced by any means without the written permission of the publisher.

Library of Congress Cataloging-in-Publication Data

Names: McCormick, Anita Louise, author.
Title: Nickel / Anita Louise McCormick.
Description: New York, NY : Enslow Publishing, LLC, [2019] | Series: Exploring the elements | Audience: Grades 5 to 8. | Includes bibliographical references and index.
Identifiers: LCCN 2017049278 | ISBN 9780766099173 (library bound) | ISBN 9780766099180 (pbk.)
Subjects: LCSH: Nickel—Juvenile literature. | Transition metals—Juvenile literature. | Chemical elements—Juvenile literature.
Classification: LCC QD181.N6 M35 2018 | DDC 546/.625—dc23
LC record available at https://lccn.loc.gov/2017049278

Printed in the United States of America

To Our Readers: We have done our best to make sure all website addresses in this book were active and appropriate when we went to press. However, the author and the publisher have no control over and assume no liability for the material available on those websites or on any websites they may link to. Any comments or suggestions can be sent by email to customerservice@enslow.com.

Portions of this book appeared in *Nickel* by Aubrey Stimola.

Photo Credits: Cover, p. 1 (chemical element symbols) Jason Winter/Shutterstock.com; cover, p. 1 (nickel ingots) AlexLMX/Shutterstock.com; p. 5 magnetix/Shutterstock.com; p. 9 farbled/Shutterstock.com; p. 10 pupunkkop/Shutterstock.com; p. 14 BlueRingMedia/Shutterstock.com; p. 18 concept w/Shutterstock.com; p. 22 Malachy666/Shutterstock.com; p. 24 Carlos Clarivan/Science Source; p. 26 TheLearningPhotographer/Shutterstock.com; p. 28 De Agostini Picture Library/Getty Images; p. 30 Bloomberg/Getty Images; p. 32 Vladyslav Danilin/Shutterstock.com; p. 35 Nuttawut Uttamaharad/Shutterstock.com; p. 40 (top) Nataliia Leontieva/Shutterstock.com; p. 40 (bottom) Rainer Fuhrmann/Shutterstock.com; p. 41 Tethys Imaging LLC/Shutterstock.com.

Contents

Introduction 4
Chapter 1: What Is the Element Nickel? 7
Chapter 2: The Periodic Table Describes Elements..... 17
Chapter 3: Where on Earth Is Nickel? 25
Chapter 4: Nickel in Our Everyday Lives 29
Chapter 5: Nickel in Living Things 39

Glossary 43
Further Reading 45
Bibliography 46
Index 48

Introduction

When you hear the word "nickel," you probably think of a five-cent coin. While "nickels" do contain a high percentage of the element nickel, and coins that contain nickel have been around for thousands of years, money is only one use for this versatile metal.

Throughout recorded history, the element nickel (Ni) and the alloys, or mixtures of elements, that can be made from nickel, has been used in many cultures around the world. Thousands of years before modern scientists identified the elements and their unique properties, people that worked with metal experimented with nickel and discovered useful applications for it. Long before it was officially identified as a unique substance, nickel (Ni) played an important role in many ancient cultures around the world.

Much of the nickel used throughout recorded history was mixed with other metals or elements. Archaeologists estimate that around

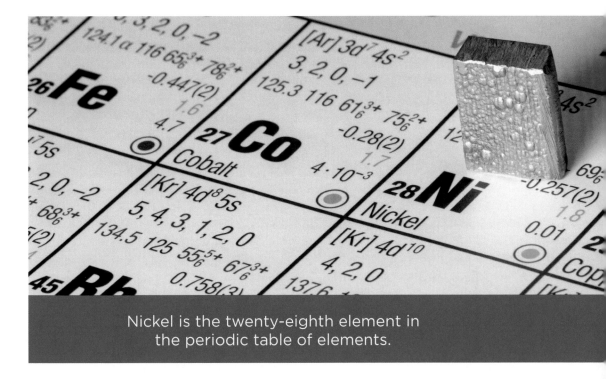

Nickel is the twenty-eighth element in the periodic table of elements.

3500 BCE, Syrian metalsmiths had learned how to make objects from bronze, which was a mixture of nickel and other metals.

China was another nation where nickel came into use. Archeologists estimate that between the years 1700 and 1400 BCE, ancient Chinese metal workers were able to develop a shiny, strong material that they called *pai-t'ung*, which translates into "white copper." They used this metal for many things, including weapons, coins, and artwork. Later, pai-t'ung was exported to the Middle East and parts of Europe. Scientists now believe that pai-t'ung was likely a mixture of nickel, zinc, copper and possibly other metals.

During the Middle Ages, metal workers in Europe often used compounds of nickel and other metals to make suits of armor, swords, and tools. Ancient Peruvians discovered a shiny, corrosion-resistant metal they believed was a different type of silver (Ag). But many scientists think it was probably nickel or a mixture of nickel and other metals.

Nickel also had uses outside of the standard applications for tools, coins, and weapons. When mixed with other elements, nickel can add color to glassware. In Germany, a nickel-containing mineral was used to color glass green.

As scientists and metal workers gradually developed more sophisticated equipment, they learned more about nickel and how its properties, alone or combined with other elements, could be used in even more innovative ways.

When electroplating was discovered in England around 1844, demand for nickel rose, as nickel was found to be a good quality metal base for silver electroplating. Not long after that, nickel was also found to be valuable as a protective, corrosion-resistant coating when it was electroplated over other metals.

Today, nickel is used in many high-tech devices that make our modern society possible. Many important components used in modern electronics contain nickel. It is used in solar energy cells that provides free electricity from the sun. Nickel is used in batteries of hybrid cars, power tools, and other devices that can be recharged. Rechargeable solar batteries that contain nickel are even used to help power spacecraft that fly to the outer reaches of the solar system and beyond to bring scientists new information about the universe we live in.

1

What Is the Element Nickel?

Nickel is a very common metal in our society. It is a hard, silvery white metal that is extremely shiny and malleable, meaning that it can easily be shaped and molded. Nickel is also ductile, meaning it can be stretched into thin sheets or wires. It can tolerate very high and very low temperatures. Nickel does not rust, and is one of only three naturally magnetic elements. As we will see, these qualities make nickel an extremely important element.

Nickel is the twenty-fourth most abundant element in Earth's crust. It can be found in all soils, in the ocean floors, and in ocean water itself. Nickel is also released during volcanic eruptions. The largest source of nickel on the Earth is buried deep in the planet's molten core. Scientists believe that nickel makes up approximately

7 to 10 percent of the planet's core. It is unlikely that deposits of nickel in the Earth's core can ever be used by humans because it is many miles beneath the Earth's surface.

How Nickel Was Discovered

In the 1700s, copper miners in the Saxony region of Germany came across a strange substance that was slightly lighter in color than copper (Cu). Unlike copper, this mysterious ore turned a bright silver color when refined, rather than reddish brown. Also unlike copper, it was extremely hard. Because no one had found a use for this uncooperative material other than to give glass a greenish hue, copper miners believed that the devil had planted the substance there to deceive them in their search for real copper. As a result, it became known as *Kupfernickel,* or "the devil's copper."

Nearly fifty years later, the strange substance that gave the ore its unique characteristics was finally identified. In 1751, mineralogist Baron Axel Fredrik Cronstedt was investigating a mineral called niccolite that was found in a mine in Sweden. Cronstedt was expecting to extract copper from this interesting ore. But his experiments repeatedly resulted in the extraction of a hard, shiny, whitish metal instead. Realizing that further attempts to extract copper from Kupfernickel would be fruitless, the frustrated Cronstedt named this new metal "nickel," the Swedish word for "Old Nick," or the devil.

Nickel is hard and silvery. What the Chinese called "white copper" was probably nickel.

At first, the chemists of the time were unwilling to accept that a new element had been discovered. They were sure that nickel was really a mixture of many other substances that were already known, such as iron (Fe), cobalt (Co), arsenic (As), and copper. It wasn't until 1775 that another Swedish chemist, Torbern Bergman, solved this mystery. When working with niccolite, he extracted a sample of nickel that was so pure that it proved beyond a shadow of a doubt that the new metal was not a combination of other substances, but was instead its own unique element.

Elements Make up Everything

An element is something that is made of only one kind of atom. When Cronstedt attempted to extract copper from niccolite, his experiments repeatedly yielded the strange, shiny, white metal from which he could extract nothing else. This inability to be broken down into simpler components is what makes nickel an element. Each element is made up of its own special atoms. For example, the element nickel is made up entirely of nickel atoms, just as the element iron is made up entirely of iron atoms.

Because they are the most basic of all substances, elements are the building blocks of everything in the universe. Elements can join together in many different ways to create molecules. All the things

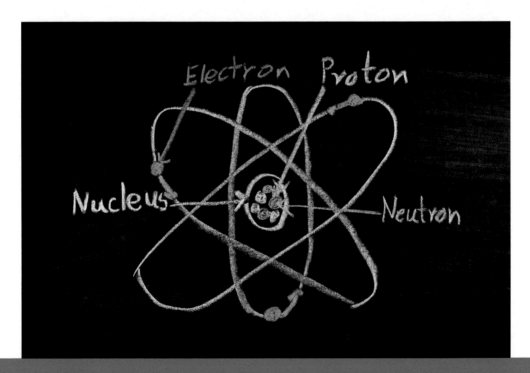

Atoms are made up of protons, neutrons, and electrons.

you see around you are made up of these particles. The air you breathe, the water you drink, even the book you are reading are all composed of a combination of various elements.

Just as puzzle pieces can be put together to make a picture, elements combine to make the range of substances in the universe by fitting together in specific ways. Different substances are made depending upon which elements fit together. For example, when an atom of the element oxygen (O) combines with two atoms of the element hydrogen (H), a water molecule is formed. Substances like water that are made up of more than one type of atom are called compounds.

Understanding Subatomic Particles

Atoms are made up of protons, neutrons, and electrons. These are known as subatomic particles. Protons and neutrons are found in the atom's center, or nucleus. Because protons have a positive electrical charge and neutrons have no charge at all, the nucleus of an atom is always positively charged. Electrons circulate around the nucleus, filling most of the space in the atom. The electrons are organized in shells. An atom can have several electron shells, and each shell can hold a certain number of electrons.

Because electrons have negative electrical charges, they are drawn to the positive charges of the protons in the atom's nucleus. This attraction is similar to the way a magnet is attracted to a refrigerator door. The attraction between electrons and protons, known as electromagnetic force, is what holds atoms together.

Since electrons have negative charges, they tend to stay as far away from other electrons as possible. The number of negatively charged electrons in an atom always matches the number of positively charged protons (the atomic number). As a result, electrons and protons cancel each other out, leaving the atom with no overall charge.

Metals, Metalloid, and Nonmetal Elements

Over three quarters of the elements known to man, including nickel, are metals. The other elements fall into one of two categories. Some, such as germanium (Ge), boron (B), and silicon (Si), are known as semimetals, or metalloids. There are seven generally accepted naturally occurring semimetals. The remaining elements are nonmetals, such as carbon (C), sulfur (S), and iodine (I). There are only sixteen naturally occurring nonmetals. However, nonmetal elements are more abundant. Two nonmetal elements, helium and hydrogen, make up a significant percentage of the universe. Oxygen, which is also a nonmetal, makes up nearly half of the Earth's atmosphere and water supply.

At Closer Look at Nickel

The number of electrons an atom has and how they are arranged in shells are important in determining how the atom will behave. Generally speaking, when an atom has a filled outer shell of electrons, it is stable and does not react with other atoms in its vicinity. When the outermost shell is not full, the atom is unstable and therefore reactive. In order to fill its outermost shell, an unstable atom will react with other atoms to either borrow or share electrons.

The number in the upper left-hand corner of nickel's square on the periodic table is its atomic number, or the number of protons in its nucleus. The number in the upper right-hand corner is its atomic weight, or the average mass of the element's isotopes. Isotopes are atoms of an element that have the same number of protons in their nucleus as other atoms of the same element, but a different number of neutrons. The mass of each isotope is the sum of all protons and neutrons in its nucleus. If a nickel isotope has twenty-eight protons and thirty-six neutrons, its atomic number is twenty-eight, and it has an atomic weight of sixty-four.

Because we know that a nickel atom has twenty-eight protons, we also know it has twenty-eight electrons. Generally,

The Discovery of Subatomic Particles

The first person to discover the existence of subatomic particles was physicist J. J. Thomson, who identified the electron in 1897. Thomson knew that atoms must also have a positive charge to balance out the negative charge of electrons. This positive charge was supplied by protons, which were identified in 1919 by the physicist Ernest Rutherford. Neutrons were discovered even later, in 1932, by the physicist James Chadwick.

When an atom loses electrons, it is no longer electrically neutral. When a nickel atom loses its outer two electrons, it becomes a nickel ion with a charge of +2. This is because the nickel ion now has two more positively charged protons than it does negatively charged electrons. Likewise, if a single atom should get the two electrons lost by nickel, it would become an ion with a -2 charge because it now has two more negatively charged electrons than positively charged protons. Because these two ions have opposite charges, they are attracted to each other and an ionic bond forms between them. Together, they are considered an ionic compound. Although a nickel atom can exist as an ion with charges ranging from -1 to +4, its most stable ion form is +2.

Because nickel atoms have an inner electron shell that is not quite full, they can form a great number of different compounds. These compounds are generally quite complex and enable nickel to play very important roles in the world.

2

The Periodic Table Describes Elements

At the present time, the periodic table lists a total of 118 elements. Ninety of these elements occur naturally. Scientists have created the remaining elements in laboratories using specialized equipment, such as a particle accelerator.

Four new scientifically developed elements were added to the periodic table in 2015. Scientists at the Lawrence Livermore Laboratory used a partial accelerator to create elements 115, 116 and 118. Scientists at the Ripken Institute in Japan created element 113 using the same type of equipment.

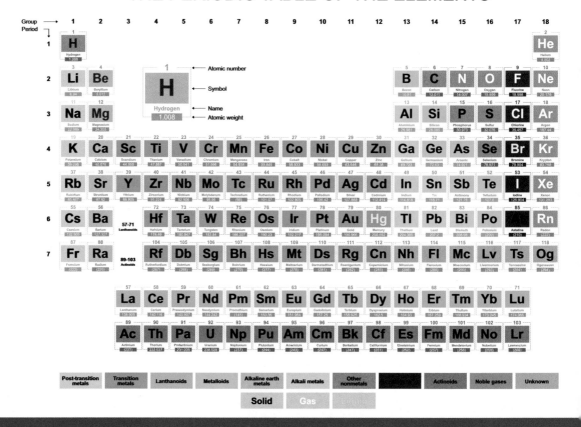

The periodic table of elements.

Scientists in most countries use the element names agreed upon by the International Union of Pure and Applied Chemistry. When new elements are discovered or created, they must decide what to call it. This process often takes time for all sides of the issue to be heard. According to the IUPAC regulations, elements can be named after mythological concepts, a mineral, a geographic location, a property they have, or a scientist.

As technology develops, it is likely that more new elements will be created. It is also possible there are still some undiscovered naturally occurring elements on Earth, as well as in the universe.

Mendeleev Categorizes the Elements

It wasn't until 1869 that a Russian chemist by the name of Dmitry Mendeleev designed a method of categorizing the elements. Mendeleev began by looking for patterns among the various elements and found that some of them shared similar chemical and physical properties, such as density and melting points. Next, he wrote down all the known characteristics of each element on note cards and tried to organize them in different ways. When the elements were arranged in horizontal rows in order of increasing atomic weight, he noticed that patterns began to emerge.

Mendeleev's organization of the elements by atomic weight did have some flaws, the largest of which was that it resulted in a lot of gaps between the known elements. The Russian chemist boldly proposed that the blanks actually represented elements that had not yet been discovered. He also predicted some of the characteristics of these yet undiscovered elements based on the location of the gaps in what he called the periodic table of the elements. (The word "periodic" refers to something that has repeating patterns.) Less than twenty years after the 1869 publication of the periodic table, scientists had filled three of the gaps with newly discovered elements. These new elements fit into the periodic table exactly where Mendeleev predicted they would.

Since its first publication, the periodic table of the elements has undergone many changes, including the discovery and addition of more than fifty elements. However, Mendeleev's method of organization continues to be the basis for our modern periodic table.

The elements are arranged in horizontal rows called periods by atomic number, or number of protons. They are also arranged into eighteen vertical columns, or groups, numbered IA through VIIA, IB through VIIIB, and 0. Just as members of a family have similarities, the elements within each group have similar chemical properties. In fact, elements in the same group are often referred to as being part of the same family.

Nickel's Place in the Periodic Table

So where does nickel fit in the periodic table? Nickel is part of group VIIIB (group 10 in some versions), which is part of a larger group of elements known as transition metals. Transition metals can be described as being typical metals. Many of them have important uses in manufacturing and industry. As we will learn, nickel is particularly important in this regard.

The Properties of Nickel

Properties are the physical and chemical characteristics that make an element unique. Like many of its neighboring transition metals, nickel is strong, hard, shiny, easy to shape, a fairly good conductor of heat and electricity, and less reactive than some other metals. Nickel also has a high melting point and a high boiling point. All elements on the periodic table can exist in one or more of three physical states, or phases: solid, liquid, and gas. Nickel is a solid at room temperature but will become liquid when heated to 2,647 degrees Fahrenheit (1,453°C). It will boil and become a gas at 5,275 degrees Fahrenheit (2,913°C).

Nickel belongs to a large family of elements known as transition metals. Transition metals are noted for the unique arrangement of their outermost electrons, which

What Qualifies as an Element?

It's not easy for a substance to be added to the list of accepted elements and earn a spot on the periodic table. In 1999, scientists at the U.S. Department of Energy thought they had created atoms of new elements 116 (ununhexium) and 118 (ununoctium) by smashing krypton ions into lead using a machine called a cyclotron. While element 116 has been re-created, no other experiments were able to duplicate the creation of element 118. As a result, in 2001, scientists took back the claim of having discovered this new element. However in 2015, scientists at the International Union of Pure and Applied Chemistry added four new elements that were created in particle accelerators to the periodic table.

First built in the early 1930s, a cyclotron is a type of particle accelerator. Particle accelerators are used to create very high-speed, high-energy beams of subatomic particles, which then collide with target atoms. The smaller particles and energy that result from these intense collisions are then analyzed in detectors. One reason scientists are interested in particle acceleration is because it provides opportunities to discover new forms of matter as well as what smaller parts make up matter.

NICKEL

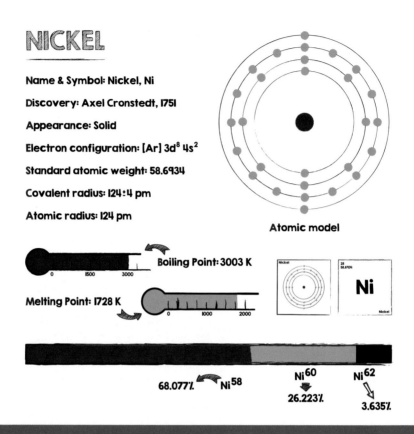

Name & Symbol: Nickel, Ni

Discovery: Axel Cronstedt, 1751

Appearance: Solid

Electron configuration: [Ar] $3d^8 4s^2$

Standard atomic weight: 58.6934

Covalent radius: 124±4 pm

Atomic radius: 124 pm

Boiling Point: 3003 K

Melting Point: 1728 K

68.077% Ni 58   Ni 60 26.223%   Ni 62 3.635%

> Nickel belongs to a large family of elements known as transition metals.

causes them to behave differently than other metals and gives them unique physical and chemical properties. Nickel is found in the first period of transition metals and is part of a smaller group known as VIIIB. Other transition metals you may be familiar with are iron, copper, zinc (Zn), and titanium (Ti).

However, nickel is one of only three elements, along with iron and cobalt, that are ferromagnetic. Ferromagnetic substances are those that can be permanently magnetized. When magnetized, they are attracted to iron and a few other metals, and make excellent materials for magnets. Nickel is also very ductile, meaning it can be stretched and hammered into thin wires and sheets. It is also quite malleable, or easily molded. Another important characteristic of nickel is its resistance to corrosion, or breaking down when exposed to oxygen in the air.

Corrosion Affects Metal

Corrosion is a chemical reaction that takes place between metal and oxygen in the presence of moisture. This reaction is called oxidation. When nickel corrodes, its surface becomes coated with a layer of oxide. This process is what causes metals like silver to tarnish, or lose their shine. In the case of nickel, this oxide layer actually protects the metal's atoms below the surface from further corrosion. Iron is a very strong metal, but it corrodes easily, resulting in the formation of rust. Rust causes metals to become brittle. To create rust resistance, iron is often coated or mixed with metals that do not corrode easily, such as nickel. In fact, the oxidation of nickel creates a coating on the surface of metals that actually protects them from further corrosion.

Extra Neutrons Create Nickel Isotopes

While most nickel atoms have twenty-eight protons and thirty neutrons, some nickel atoms have extra neutrons. These forms of the

Most nickel atoms have twenty-eight protons and thirty neutrons. Some nickel atoms have extra neutrons.

element are called nickel isotopes. Many elements, including nickel, have more than one isotope.

You can think of isotopes as different forms of the same atoms. Neutrons have mass but no electrical charge, so nickel isotopes have only slightly different physical properties, such as atomic weight. There are five stable isotopes of nickel. The official atomic weight of nickel (approximately fifty-nine atomic mass units) is the average of the mass of all five nickel isotopes.

3

Where on Earth Is Nickel?

We already know that nickel is the twenty-fourth most abundant element in Earth's crust. Scientists estimate that 7 to 10 percent of Earth's molten core is made up of nickel. But how did it get nickel there? Nickel's story goes back to the time when the Earth was in the process of being formed into a planet.

Earth is thought to have formed billions of years ago when a large number of particles drew together into a cloud. The cloud grew increasingly dense as it attracted more and more particles. Eventually, this massive cloud condensed into a liquid. Heavier substances like nickel and iron sank and collected in the center of the newly formed planet, creating its core.

Meteorites—rocks that plummet from space—
are a good source of nickel.

Much of the nickel that is not buried deep within Earth's core is thought to have arrived more recently in the form of asteroids, meteorites, and space dust.

Throughout Earth's 4.5-billion-year lifetime, the planet has been pummeled by space debris and meteorites. In fact, scientists believe that the world's biggest nickel deposit, which is located in Canada, formed when an enormous asteroid containing large amounts of nickel struck Earth almost two billion years ago. Besides containing a lot of nickel, scientists believe that the asteroid's forceful impact also resulted in volcanic eruptions. During these eruptions, a great deal of nickel was expelled from below Earth's surface.

Mining and Refining Nickel

Nickel is rarely found in nature in its pure form. Instead, it is locked away in mineral ores, such as the niccolite that frustrated German miners looking for copper. Nickel is generally found in sulfide ores called pyrrhotite and pentlandite.

It is also found in complex ores known as laterites. Before they can be used for manufacturing, all of these ores must first be extracted from the earth and then purified in refineries in order to make use of the nickel within.

A number of different methods are used to extract nickel from the earth. When nickel ores are buried deep in the ground, tunnels are dug in the earth and the nickel ore is excavated and transported back up to the surface. When nickel ores are found closer to the surface, explosives are used to blast them out. Once removed from mines, the nickel ores are sent to refineries, where pure nickel is separated from the rest of the minerals. This is precisely the task that Cronstedt undertook when he tried to break down "the devil's

Identifying Meteorites by Their Nickel Content

There are three different types of meteorites: stony, iron, and stony-iron. The percentage of nickel a meteorite contains helps distinguish which category it belongs in. The most common meteorites are of the stony variety and contain 15 to 25 percent nickel-iron. Iron meteorites, the second most common type of meteorite, may be composed almost entirely of a nickel-iron combination.

Pyrrhotite, an iron sulfide, also contains nickel.

copper" into its most basic, elemental components. Modern technology has since made this process far easier than it was for its pioneer, Cronstedt.

The nickel refining process begins by pulverizing the ores to create a fine powder. For the sulfide ores, pyrrhotite and pentlandite, this powder is mixed with water. Air bubbles are introduced into the water. The particles containing nickel stick to the air bubbles, which are then carried to the surface where they form a thick film. This film is scraped off and dried, resulting in a powder of pure nickel sulfide. Repeating this process several times ensures that as many of the nickel-containing particles as possible have been captured.

The nickel sulfide powder is then smelted—a high-temperature process that causes pure molten nickel to sink and waste products, or slag, to float to the top where they are removed. This process, too, is repeated several times, resulting in a nickel sample that is almost 100 percent pure.

For complex laterite ores, different processes are used to refine pure nickel. These multi-step processes often involve binding nickel to sulfides to make it possible to use refining methods similar to those employed for refining pure nickel from pyrrhotite and pentlandite ores.

Nickel in Our Everyday Lives

Nickel plays very important roles in our everyday lives. Even though most people are unaware of it, many items we use at home, school, and work are made of nickel and its alloys. Because of nickel's unique properties, it is used in building materials, batteries, water purification, chemical production, food preparation, machinery, household appliances, and transportation. Every time nickel is mixed with another kind of atom or molecule, the process creates an alloy, which has properties of its own.

Nickel Electroplating

Because nickel is naturally resistant to the effects of oxygen and other chemicals, thin layers of pure nickel are often applied to the

A worker separates copper and nickel in an electroplating bath. This technique is used in the manufacturing of vinyl records.

surfaces of metals that are not as resistant to corrosion. One process for accomplishing this is called electroplating.

Electroplating involves the attraction of positive nickel ions in a chemical solution to a negatively charged metal surface. The ions gain electrons from the negatively charged metal and form nickel atoms. These atoms create a uniform covering that shields the metal beneath from the corrosive effects of oxygen. Currently, scientists are even using electroplating to protect plastic surfaces. The plastic is first treated with a coat of solid metal, which is then electroplated with nickel. Electroplating is used to protect jewelry, computer disks, and kitchen and bathroom hardware.

Nickel Compounds

When the German miners unearthed copper-colored niccolite in their quest for true copper, they found that the uncooperative substance could be used to color glass green. The substance responsible for this green coloration was a compound called nickel arsenide. Similar nickel compounds are still used today in glass coloring and pottery glazes.

Rechargeable batteries require the use of an important nickel compound. Batteries harness energy released from contained chemical reactions and turn it into electrical energy that can be used to power devices like cell phones,

How to Make a Battery with Pennies and Nickels

To make a coin battery, you'll need five pennies, five nickels, soap, a salt-water solution, paper napkins, and a volt meter.

Directions:
1. After cleaning the coins with soap, soak the napkins in saltwater solution.
2. Tear the napkins into pieces slightly larger than the size of the coins, and make a "sandwich" by stacking a penny, a piece of napkin, a nickel, a piece of napkin, and so on.
3. Connect a voltmeter to the ends of the stack.

What Makes It Work?
The saltwater solution is an electrolyte. The electrolyte reacts with metals in the coins, which act as electrodes. The two coins contain different metals (copper in the penny and a nickel-copper alloy in the nickel), one of which reacts more strongly than the other. This leaves an electrical potential difference between them. In batteries, this difference is called voltage.

Rechargeable batteries use nickel and keep our landfills free of corrosive batteries.

radios, and even cars. All batteries have a positive end (cathode) and a negative end (anode) made from two different metals or metal compounds. An electrolyte solution between the cathode and anode allows chemical reactions to take place between them. This chemical reaction causes a buildup of electrons at the anode. Just as the negatively charged electrons in an atom are attracted to the positively charged protons in the atom's nucleus, the electrons at the anode are attracted to the positively charged cathode.

However, the chemical reaction occurring in the battery keeps this from happening. When a battery is inserted into a device that requires power and the device is turned on, a closed circuit, or a path for the electrical energy to travel, is created. The electrons from the anode then travel through the device in order to reach the cathode. The electrons release energy into the device in the process.

In nickel-containing batteries, the cathode contains a compound called nickel oxyhydroxide. The anode can contain either a metal hydride or the element cadmium. Nickel-containing batteries are much more efficient and convenient than other kinds of batteries because they can be recharged. In other battery types, the chemicals involved in the reactions lose their ability to supply electrons, causing the battery to die. But the chemical reactions that take place in nickel-containing batteries can be reversed, restoring the battery to its original state, chock-full of stored energy.

The element nickel has the ability to bond with large amounts of hydrogen gas to form nickel hydride. Hydrogen, when it reacts with oxygen, is a powerful source of pollution-free energy, but the reaction can be extremely dangerous if not carefully controlled and contained. Nickel and other metal hydrides could be the easiest and safest way to store hydrogen for use in energy-producing reactions. Scientists are working on ways of releasing the hydrogen bound in nickel hydrides in a way that would allow them to efficiently capture the most energy.

Nickel's Use in Steel Production

Without question, one of the most important role nickel plays in our world is in the production of steel. Steel is a very strong metal that is widely used in industry and technology. Steel is used in the construction of everything from skyscrapers to cars to medical equipment. Steel itself is not an elemental metal, however. Instead, it is a mixture of metals, or an alloy.

Alloys are valuable because they have different properties from the elements that make them up. Each element on its own cannot perform the functions that alloys of that element can.

Nickel-containing steels are resistant to corrosion and are very ductile. The use of these steels can be traced all the way back to 1889, when engineer James Riley used them to make armor plating that was both strong and heat resistant.

The properties of different kinds of nickel-containing steels depend on the combination of elements in addition to nickel that are added to iron. For example, some nickel steels are more ductile than others. These stretchy steels are used in structures that must give under tremendous pressure without breaking or snapping, such as bridges. Other nickel steels can withstand very low temperatures and are used in containers that store liquids with very low boiling points, such as nitrogen and hydrogen. There are also nickel steels prized for their extreme hardness; these are used in rocket fuel tanks.

5
Nickel in Living Things

Nickel is not just something we use for manufacturing, we also eat it in our food. Scientists that study nutrition estimate that people only require very small amounts of nickel in their diets to stay healthy. They are still unsure, however, of exactly what role nickel plays in the human body.

Scientists have conducted experiments where they observe what happens to animals that are biologically similar to human beings when they don't get enough nickel. For example, when baby chicks and rats don't get enough nickel, they sometimes have liver problems. The liver is an important organ that plays many roles, including the production of substances that help in the digestion of food. Nickel may also be involved in growth, the production of

Foods such as walnuts, almonds, oatmeal, cocoa, and soy beans have high levels of nickel.

breast milk, and helping our bodies make proteins. Foods such as walnuts, almonds, oatmeal, cocoa, soy beans, fresh and dried vegetables, peas, beans, and tea leaves contain large amounts of nickel.

Nickel has important functions in many plants. Nickel helps some plants absorb iron, it helps others to break down certain chemicals, and it may be necessary in the sprouting of seeds. Tea has more than twice the amount of nickel in its leaves than most other plants, and nickel also plays various roles in fungi and bacteria.

Nickel Allergies

Even though humans need nickel, too much exposure to nickel can cause problems. Sometimes people have allergic reactions to nickel from wearing inexpensive

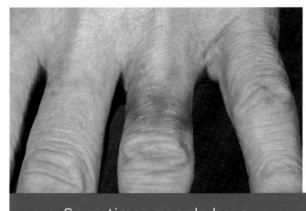

Sometimes people have allergic reactions to nickel used in inexpensive jewelry.

jewelry that is electroplated with nickel or nickel alloys. This allergy can develop over time, even after years of wearing the same kind of jewelry. This gradual process is called sensitizing. An allergic reaction to nickel generally appears as a reddish, bumpy, and sometimes blistery rash that can be very itchy. It is often made worse by sweating. A reaction caused by something that irritates the skin is called contact dermatitis. The rash generally heals when the nickel-containing item is no longer worn. People with nickel allergies should wear only jewelry that is silver, gold (Au), or stainless steel, and they should avoid wearing items plated with nickel alloys against their skin.

Nickel's Use in Coins

The United States was not the first country to use nickel in its coins. In fact, nickel was first used to make coins in 1860 in Belgium, and the first coins made from pure nickel were made in 1881.

Nickel was first used to make coins in 1860 in Belgium.

The first five-cent coins in the US were not made of nickel. In the United States during the mid-1800s, five-cent coins called half-dimes were minted from silver. After the American Civil War, however, there was a shortage of silver, which resulted in the decision to use a combination of copper and nickel for these coins. Copper alone bent too much and corroded easily. The same copper-nickel alloy is used in all US coins today, but in different amounts. For example, if you look at the edge of a US quarter or dime, you'll notice layers of different colors. That's because only the faces of the coins are made from a copper-nickel alloy; the insides are made of pure copper.

In 2002, researchers in Europe discovered that some of the new coins produced by the European Union caused allergic reactions when held in the hands of nickel-sensitive people. The coins in question contained two distinct nickel alloys. Electrolytes in sweat from the hands of this small minority of nickel-sensitive people caused the coins to corrode faster than usual, releasing more nickel into the skin than other coins do.

Throughout this book, we have learned that nickel plays many important roles. Nickel gives us rechargeable batteries, which can be reused rather than thrown away. It makes flying around the world and even into space safer and easier. It makes bridges and buildings strong, and it even makes some important chemical reactions happen faster. Everywhere you go, you can bet that nickel is either right there on the surface or hiding just beneath it.

Glossary

alloy A mixture of two or more metals, or a metal and a nonmetal.

anode The negative electrode of a battery.

atomic mass unit A unit of mass that is approximately equivalent to one atom of hydrogen.

atomic number The number of protons in the nucleus of an atom.

atomic weight The average mass of all isotopes of an element.

battery A source of electrical potential energy made up of two or more electrochemical cells.

cathode The positive electrode of a battery.

corrosion The process by which the surface of a metal is worn away by oxygen and chemicals.

cyclotron A type of particle accelerator in which charged particles are accelerated while confined to a circular path by a magnetic field.

ductile A term used to describe substances that can be stretched and hammered into thin wires and sheets.

electrode A conductor through which an electrical current enters or leaves.

electroplating The process of covering an object with a thin layer of metal using electrolysis.

element A substance made up of one type of atom that cannot be broken down into simpler substances.

ferromagnetic A term used to describe metals that are easily magnetized.

malleable A term used to describe metals that are easily shaped and molded.

meteorite A meteor that has fallen through the atmosphere and landed on Earth.

mineral Any of the naturally occurring compounds of which rocks and ores are made.

molecule The smallest particle of an element, consisting of one or more atoms, that exists on its own and still maintains its properties.

niccolite A mineral that consists of 43.9 percent nickel and 56.1 percent arsenic.

ore A mineral from which useful substances like metals can be extracted.

refining The process of removing minor impurities from metal ores.

smelting The process of extracting a metal from an ore by repeatedly heating it to high temperatures and removing impurities.

turbine A machine with blades attached to a central shaft. The pressure of water or steam on these blades causes the turbine to spin.

Further Reading

Books

Baby Professor. *The Periodic Table of Elements – Post Transition Metals, Metalloids and Nonmetals*. Children's Chemistry Book, Baby Professor, 2017.

Callery, Sean, and Miranda Smith. *The Periodic Table*. New York: Scholastic Nonfiction, 2017.

Dingle, Adrian, and Simon Basher. *Basher Science: The Complete Periodic Table: All the Elements with Style!* New York: Kingfisher, 2015.

DK. *The Elements Book: A Visual Encyclopedia of the Periodic Table*. New York: DK Children, 2017.

Green, Joey. *The Electric Pickle: 50 Experiments From the Periodic Table, From Aluminum to Zinc*. Chicago, IL: Chicago Review Press, 2017.

Websites

Chemistry for Kids
www.ducksters.com/science/chemistry/nickel.php
Explore nickel and other elements on the periodic table.

Element Facts
www.thoughtco.com/interesting-nickel-element-facts-3858573
Ten keen facts about nickel, including photos!

KidzScience
wiki.kidzsearch.com/wiki/Nickel
An awesome collection of nickel facts, including photos of its various forms.

Bibliography

Atkins, P. W. *The Periodic Kingdom: A Journey into the Land of the Chemical Elements*. New York, NY: Basic Books, 1997.

Emsley, John. *Nature's Building Blocks: An A–Z Guide to the Elements*. New York, NY: Oxford University Press, 2011.

Heiserman, David L. *Exploring Chemical Elements and Their Compounds*. New York, NY: McGraw-Hill, 1991.

Helmenstine, Anne Marie Ph.D. "Nickel Facts," *ThoughtCo*. June 21, 2017. Retrieved October 7, 2017. https://www.thoughtco.com/nickel-facts-606565.

Kluge, Jeffrey. "How Four New Elements Got Seats at the Periodic Table," *Time Magazine*. January 4, 2016, Retrieved October 10, 2017. http://time.com/4165979/periodic-table-new-elements.

Mayo Clinic Staff, "Nickel Allergy," Retrieved October 7, 2017. http://www.mayoclinic.org/diseases-conditions/nickel-allergy/symptoms-causes/dxc-20267456.

Morrison, R. T., and R. N. Boyd. *Organic Chemistry*. 6th ed. San Francisco, CA: Benjamin Cummings, 1992.

Pedersen, Traci. "Facts about Nickel," *Live Science*. September 23, 2016. Retrieved October 7, 2017. https://www.livescience.com/29327-nickel.html.

BIBLIOGRAPHY

Rogers, Kirsteen, et al. *The Usborne Internet-Linked Science Encyclopedia.* London, England: Usborne Publishing, 2003.

Sparrow, Giles. *Nickel.* New York, NY: Benchmark Books, 2005.

US Geological Service. "Nickel: Statistics and Information," Retreived October 10, 2017. https://minerals.usgs.gov/minerals/pubs/commodity/nickel.

Woodford, Chris. "Nickel," *Explainthatstuff*! August 5, 2017. Retrieved October 10, 2017. http://www.explainthatstuff.com/nickel.html.

Index

A
alloys, 4, 29, 31, 34–38, 41–42
atomic weight, 13, 19, 24
atoms, 10–16, 19–21, 23–24, 29–30, 32, 38

B
Bergman, Torbern, 9

C
compounds, 11, 15, 31–33
Cronstedt, Baron Axel Fredrik, 8, 10, 27–28

E
elements, 4, 6, 7, 10–12, 15, 17–22, 24, 34–38

I
ions, 16
isotope, 13, 23, 24

M
Mendeleev, Dmitry, 19–20
monel, 36

N
nickel
   alloys, 35–38
   biological role, 39–42
   discovery of, 4–6, 8–9
   properties, 20–24
   sources, 25–28
   uses, 29–38

P
periodic table, 13, 17, 19–21